THE NEWTOWN STORY

ONE COMMUNITY'S FIGHT FOR ENVIRONMENTAL JUSTICE

By Ellen Griffith Spears

Photographs by Michael A. Schwarz

PUBLISHED BY
THE CENTER FOR DEMOCRATIC RENEWAL
&
THE NEWTOWN FLORIST CLUB

All proceeds from this book go to
The Newtown Florist Club

ISBN I-0-96590305-0-5

Published by the Center for Democratic Renewal & The Newtown Florist Club
Atlanta, Georgia, United States of America

The Newtown Florist Club
1067 Desota Street
Gainesville, Georgia 30501

Book design by Donald R. Winslow, San Francisco
Maps by Roderic Johnson

Ruby Wilkins

Dedicated to

Ruby Wilkins, Ruth Cantrell, Lee Whelchel, Roland Waller and
all those who walk with us as we continue the fight for our community

And the children, who represent our greatest hope for the future

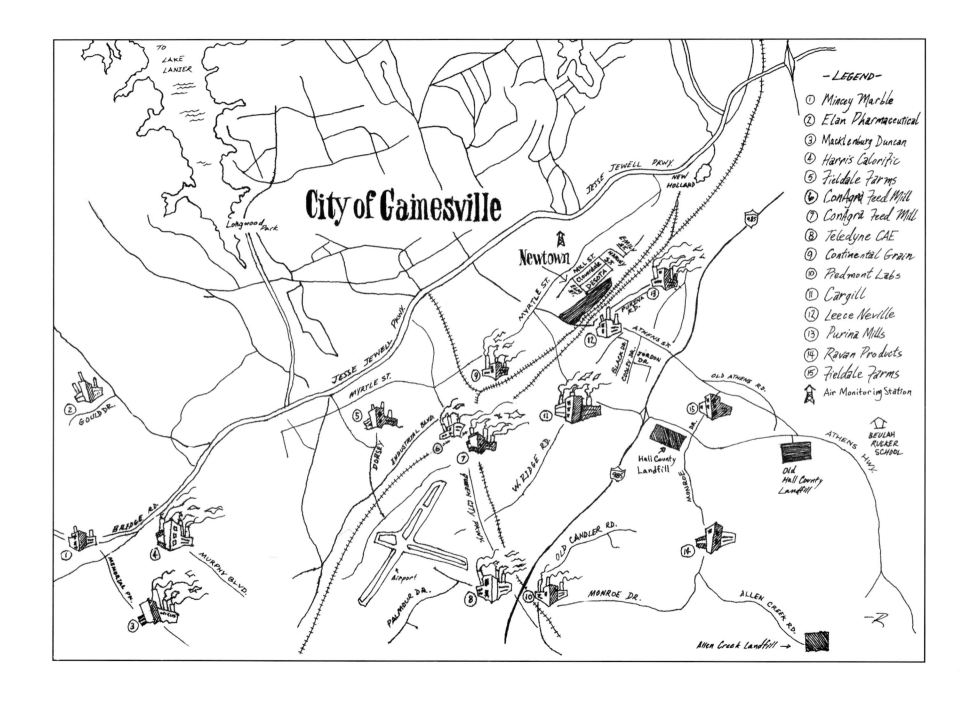

City of Gainesville

Newtown

- LEGEND -

1. Mincey Marble
2. Elan Pharmaceutical
3. Macklenburg Duncan
4. Harris Calorific
5. Fieldale Farms
6. ConAgra Feed Mill
7. ConAgra Feed Mill
8. Teledyne CAE
9. Continental Grain
10. Piedmont Labs
11. Cargill
12. Leece Neville
13. Purina Mills
14. Ravan Products
15. Fieldale Farms

Air Monitoring Station

TO LAKE LANIER

Longwood Park

JESSE JEWELL PKWY.

NEW HOLLAND

485

GOULD DR.

JESSE JEWELL PKWY.

MYRTLE ST.

MYRTLE ST.

MILL ST.
Cloverdale
DESOTA
EMILY ST.
HARRIS ST.

PURINA RD.

ATHENS ST.

BLACK DR.
COOLEY DR.
JORDON DR.

OLD ATHENS RD.

ATHENS HWY.

BEULAH RUCKER SCHOOL

BRIDGE RD.

DORSEY

INDUSTRIAL BLVD.

W. RIDGE RD.

MONROE DR.

MEMORIAL PK.

MURPHY BLVD.

QUEEN CITY PKWY.

Airport

PALMOUR DR.

OLD CANDLER RD.

MONROE DR.

Hall County Landfill

Old Hall County Landfill

985

ALLEN CREEK RD.

Allen Creek Landfill →

-R

A hearse waits on Desota Street for the funeral of a Newtown resident.

"If there is anything you all can help us do to save this community... I'd be grateful."

–Mozetta Whelchel

Newtown Florist Club members
Faye Bush, Mozetta Whelchel, Geraldine Collins

Introduction

Environmental contamination is an invisible killer that steals in quietly. We want the world to know what is happening in Newtown. We hope our story will educate others about the dangers of toxic contamination, so more men, women and children won't suffer and die needlessly.

In the spring of 1990, we realized many of our loved ones were dying from the same kinds of cancers. To get a closer look at the problem, several members of the Newtown Florist Club conducted a door-to-door survey. We were shocked to learn that as many as 18 people, most of them on the same street, had died or were suffering from some form of throat, lung or colon cancer, lupus or chemically-induced heart disease. Today, the numbers are even higher.

During our information gathering, we learned that not only was the Newtown community built on top of a landfill, it also is located in an "industrial fallout zone."

We organized several "Toxic Tours" of the neighborhood. In this report, we want you to see the victims and hear their stories. We also believe it is important for you to hear from professionals in the field who know about the impact toxic chemicals can have on health.

Many people have told us that the problems we face in Newtown are so complex that even they don't know how to figure them out. When we started, neither did we. But every step of the way and every day we have made progress. Putting this book in your hands is part of that progress.

To the victims and their families we want to say that we are sorry that it took so long – that so many had to suffer needlessly before we realized we should have taken action.

The Newtown Story: One Community's Fight for Environmental Justice is our story – a story we are sharing in the hope that one day people will not die from exposure to toxic pollutants because of the color of their skin, because of where they live, or because they are poor.

We invite you to join us and share our struggle.

Faye Bush, President
Newtown Florist Club

Rose Johnson, National Program Director
Center for Democratic Renewal

In the rising heat of a Georgia spring morning, community leader Rose Johnson guides several dozen visitors on a somber mission – the first Toxic Tour of Newtown. The date is May 7, 1993. This African American community on the south side of Gainesville, fifty-five miles north of Atlanta, reports unexplained high rates of throat and mouth cancers, excessive cases of the immune-system disease lupus, and a variety of respiratory ailments. Too many people have died.

The visitors from the Racial Justice Working Group of the National Council of Churches, along with city and state officials, follow Johnson "door-to-door, neighbor-to-neighbor" as she places black ribbons at homes where residents are sick with cancer or lupus or where a family member has died.

The observers sense firsthand the sharp contrast between the Newtown environment – where the acrid odor of toxic industry and scrap yard presses in on the little park and well-tended homes – and the flourishing green lawns and flowering trees of Longwood Park and the generous houses on the north side of town where most whites live.

On this already overheating day, an intolerable picture emerges. The seventy-five homes in Newtown, built atop an old dump, are surrounded by thirteen toxic industries, two identified potentially hazardous sites, numerous hazardous waste generators, and a rat-infested junkyard. This neighborhood next to the railroad tracks is encircled by so many toxic sites that the local paper called it "an industrial fallout zone."

As guide, Johnson represents the younger generation of leaders of the "one group that didn't mind tackling anything" – Newtown Florist Club. Founded nearly fifty years ago to pool funds for funeral wreaths, today the Club is working aggressively to uncover the environmental links to the diseases affecting residents – and to halt the toxic assault on the community.

Johnson, who played along Desota Street as a child in the 1960's, introduces the Club members and other residents one by one. Faye Bush survived her own battle with heart disease and lupus to steer the Club in key phases of the fight; her daughter Jackie Mize, fondly remembers growing up in the close-knit community. Bush's sister Mozetta Whelchel lost her teen-aged daughter, Moselee, and son, Deotris, to lupus. Her husband, Lee, died of cancer after a lifetime of work in the starter motor plant across the railroad tracks from their home.

Geraldine Collins raises her voice in protest despite throat cancer. Mae Catherine Wilmont, who lost her nursing job after she contracted lupus, is coming forward as a leader in the Club. Jerry Castleberry watched his mother die of lupus and now faces the daily pain of the disease himself.

Others have passed on – people like Ruby Wilkins, who hosted vote-seeking white politicians at her dinner table and counseled a generation of neighborhood youth who played basketball in her side yard.

Rose Johnson leads a "toxic tour" of the Newtown community.

Ruth Cantrell, who survived the deadliest disaster in Gainesville history, the 1936 tornado, was too ill from battling cancer to leave her oxygen tank to greet those gathered for the Toxic Tour on that day in 1993. She has since died. Neighbor Roland Waller, who worked with Lee Whelchel over vats of toxic chemicals at the starter motor plant, had to give up his vegetable plot because the ground was so contaminated. He died, too, of congestive heart failure.

"I thought I was immune to the pain of it," says Johnson, "but I don't guess I am."

She explains that the Florist Club began its work in 1950 with a simple, traditional mission: to care for the sick and comfort families as they buried their dead. The Club's concern for the health of individuals led inevitably to action to protect the health of the entire community.

Tested during segregation, shaped by the civil rights movement, the women and men of Newtown came forward to resist every indignity faced by Gainesville's black residents. Braving personal illness and tragedy, members have organized an endless variety of community-building, youth-developing, race-uplifting strategies. When the Klan tried to march near Newtown, the Club backed them down. When the city's election system undermined black voting strength, the Club took them to court. When awareness of the environmental threat emerged, the Club tackled toxic polluters, demanding changes from industry and pressing city, state, and federal officials for results.

The environmental fight has stretched the Club in new directions, into the underdeveloped science of

A confederate statue remained standing in Gainesville's square while the city lay devastated after the 1936 tornado. Photo courtesy of The Times (Gainesville).

ecological cause and effect, seeking toxic sources and health treatments for the ailments that plague the neighbors. When a state health survey blamed high cancer rates on residents' "lifestyle," volunteer eco-experts and health and science professionals helped members research the local toxic profile and seek environmental links. Proving a connection is difficult, but experts agree it's still fair to limit exposure. None of the work has been easy: corralling the efforts of volunteers, battling the insensitivity of a maze of bureaucrats, bringing pioneering legal claims, and garnering the attention of those who could effect a change.

An old yellow school bus transports the invited guests through the noisy heat, past the industrial sites

Railroad tracks and the industrial zone border Newtown homes. Desota Street runs parallel to the rails.

Catherine Earls lives on Desota Street.

that surround Newtown. The tour winds through the rapidly growing city of Gainesville, the largest industrial hub north of Atlanta in the state, the last urban center before the rural farmland and mountains heading north, a gateway between the future and the past. Combined with suburbanized Hall County, the area is home to more than 111,000, making it the fifth largest metropolitan area in the state. The city's business leaders have been remarkably successful in recruiting industry. Many factories depend on the area's agricultural roots, chicken processing plants and animal feed manufacturers. Gainesville boasts the slogan, "chicken capital of the world"; three poultry processors are located near Newtown. Animal food giants Ralston-Purina and Cargill operate major feed mills within earshot. Grain dust, runoff, sewage, air pollution, and groundwater contamination create problems that extend beyond Newtown. Observers learn that the nearby streams fill the Chattahoochee River and Lake Lanier, the popular resort center which is central to Gainesville's economy. Toxins endanger other black neighborhoods on the south side, where two emergency evacuations in 1995 sent dozens to the hospital and brought residents to challenge Cargill with releasing dangerous hexane emissions.

The tour participants disembark at Bethel AME Church, feeling the weight of the evidence of life in a toxic zone. Newtown's perseverance against these environmental odds has become a symbol for other communities across the South. Partial victories along the way – and deep faith – have sustained the fight, though much work remains.

But Newtown's story really begins sixty years ago.

Shirley Browner leads Newtown Florist Club members into church for the dedication ceremony for the monument to Newtown residents who have died.

"It's coming a storm today."

New Town it was called then. Seventy-five families moved into the first houses, built atop debris and wreckage from the 1936 tornado that destroyed eight hundred homes in Gainesville.

"Miss Ruth" Cantrell moved into New Town when it was built. She was twenty years old and living on Copeland Street in Gainesville when the tornado slammed into town on a Monday morning, April 6, 1936.

"It was doing everything that morning. It rained a little bit and snowed a little bit and the sun come out a little bit. My Mama told me – I lived next door to her – 'If the clouds don't look right, don't let the children go to school. I feel like it's coming a storm today.'"

Clocks in Gainesville stopped at 8:28 a.m. "God wiped up this place in three minutes," Cantrell recalled. "It flattened our house. My sister's baby, he died that night. I don't know whether she squeezed him to death or whether something hit him. My sister, she died after that. They couldn't find that she was hurt. She was just scared to death." The tornado's dual funnels ripped from the foot of West Washington Street across the heart of the business district before lifting past the New Holland Mill, more than two-and-a-half miles away. In its shattered wake lay more than two hundred dead and 950 injured. A bronze Confederate soldier in the town square stood untouched, but around the statue, the city was devastated.

A sense that the tornado represented retributive justice for the city's racial sins remains powerful among

From left: In 1938, Geraldine Hops, 12, Minnie Nell Hops (Moorehead Moody), 4, and Willene Hops Robinson, 14, in front of their home at 1070 Desota Street. Photo courtesy Pat Moorehead.

a few community elders. The tornado's unusual trail figures in a long told story, undocumented, that the twin twisters followed the route of a lynching, a path along which two black men accused of rape were dragged behind a wagon. But the twin funnels were no respecter of color in cutting their swath; the destruction fell at least as severely on the black community as the white.

"More than 600 residences were swept away, not including those blasted out of the negro district," reported The Atlanta Constitution on April 7. "The

Amanda Keith carries flowers at the funeral of a Newtown resident.

Geraldine Collins in her Newtown home.

negro section of the city was leveled by the blow. Hardly a home remained standing."

Specific news coverage of the impact on Gainesville's Negro community, which numbered 2,051 people out of 8,624 city residents according to the 1930 Census, is scant. The Constitution reported that "Negroes who escaped death or injury wandered dazedly through the wreckage. Some tried to dig out victims from the debris and others appeared too stunned to know what happened."

The nearest member paper of The Associated Negro Press, the Atlanta Daily World, reported "there are approximately 200 Negro families in Gainesville, whose destroyed homes carried no tornado insurance." Counting renters, the Daily World reported roughly 300 homeless Negro families.

Attention was already turning to rebuilding four days later, when President Franklin Roosevelt, on his way back to Washington from Warm Springs, made a thirty-minute train stop to survey the damage and offer federal help to the reconstruction effort. But the dislocation persisted.

After the storm, "Mrs. Watson took us in, and we were from pillar to post," Ruth Cantrell recalled. "We stayed out at Uncle Rich's across the river for a while. We was trying to stay together ... ten of us ... Yes, we had a time, and then they started building over here, building all these little houses."

"It used to be a trash pile over here," Cantrell explained. Benjamin Rucker, who died in 1996, recalled hunting snakes at the old dump. The storm had destroyed Gainesville city records, so no written

Benjamin Rucker, who died in 1996.

documents prove what older residents remember: that the New Town houses were built on a landfill.

Using federal Reconstruction Finance Corporation loans, Daniel Construction Company of Anderson, South Carolina built seventy-five units on Desota and Cloverdale Streets to house black families displaced by the tornado. Each 28- by 22-foot house had four rooms, with a fireplace and cold running water. "No bathtub, no hot water," said Cantrell.

Completed in March 1937, the little dwellings were advertised by a non-profit organization, Gainesville Replacement Homes, Inc., "for colored purchasers" who suffered damage in the tornado. Family incomes of those who applied under a ten-year rent-to-purchase plan averaged less than $48 a month.

The houses were new, but the segregation and discrimination were all too familiar. The fifty houses built in the "white new town" on North Main Street

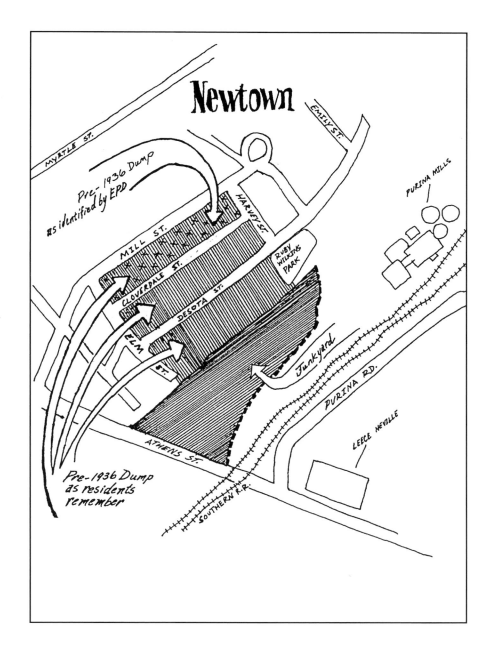

Industry looms over Desota Street homes.

were larger, and the white families who applied to move in had incomes almost three times higher, about $129 a month.

For residents of New Town, the modest incomes came from jobs the men held as laborers, janitors, kitchen helpers and bellmen. The primary paying work for black women was in white homes. "I was working at one house," Cantrell recalled. "After dinner about two o'clock, I'd leave there and go out on Northside Drive and iron, and come home and start my supper, go babysitting at night 'til about two, get up the next morning and be at work at seven."

Cantrell's memories of World War II reflect how hardships were felt more sharply by black citizens. Ration stamps were issued to all. But there were no stores in New Town, so Cantrell and others had to walk the half mile into downtown Gainesville to shop. "I went to town. I heard that they would allow some lard and sugar that day, so I kept the book of stamps and went up there and stood around." The grocer was dispensing rationed items, wrapped in newspaper. "So, I went up there and asked the man could I have some – we called it lard – didn't call it Crisco."

"'Do you still have some lard today?'"

"'We had some,' he said, 'but it's all gone.'"

"That's the way they did us," said Cantrell. "And I said, 'I don't have to never spend another penny in that store.'"

Before the war ended, Gainesville Replacement Homes went under, foreclosed by the federal Reconstruction Finance Corporation. Although the project has been publicized as a ten-year rent-to-own

Mozetta Whelchel holds a photograph of her daughter, Moselee, who died from lupus.

Ruth Cantrell moved into Newtown when it was built following the 1936 Gainesville tornado.

plan, fewer than ten years had passed and the deeds for the little houses had not been transferred to the families who lived in them. After the war, Interstate Accident and Life Insurance Company of Chattanooga, Tennessee, bought the tract of land that included New Town. The insurance company sold several of the homes in 1950 for about $2,000 each to the black families living in them or to others migrating from smaller communities around Georgia, places such as Jackson, Eatonton, Martin, and Grantville, to find work or go to school.

Nearby Athens Street had been the heart of black Gainesville since the 1920's. Sara Nash remembers a thriving community life in 1950, going to "Chamblee's Drug Store, Clearview Restaurant, Carl's Spoon, Carter's Shoe Shop, Greenlee's Funeral Home, Mr. Morgan's little cleaning place, Daddy Poole's, Asberry's barbershop." The all-black Gainesville Eagles baseball team played at cab driver "Doc" Harrison's field out Athens Highway.

"Through a child's eye it was nice," says Jackie Mize, daughter of Newtown Florist Club President Faye Bush. "It was just a small community. We all knew each other, and we all grew up over here. You know, it was a family community, where families stayed. We all grew up until we were eighteen in the same community and most families still live here that I grew up with. Their grand-kids, my kids and now my grand-kids."

New Town had been planted along the Gainesville-Midland and Southern railroads. As children, Mize and her friends played in the pulpwood yard along the railroad tracks that ran behind the houses on the east side of the dirt road that was Desota Street. They hiked past the clearing behind Emily Street along the tracks to the old Indian cave, rumored to be among those abandoned when the Cherokees were driven from their historic lands during the 1830's Trail of Tears forced migration.

More than one historian suggests the lure of the limestone caves lingered from another era as well. "Caves around Hall County might have been resting places for Georgia slaves on their way to freedom" – stops on a quite different railroad – wrote historian Linda Gail Housch, whose father pastored a church in Gainesville, in a 1969 Gainesville Daily Times article. "The limestone caves of the region served a useful purpose," stated Wilbur Siebert in his book The Underground Railroad from Slavery to Freedom.

The steam trains that rolled past New Town along the Gainesville-Midland route had since 1871 established Gainesville as an important trading hub for surrounding farms. The rich north Georgia Piedmont area, in the foothills of the southern Appalachian mountains and close to the Chattahoochee River, attracted travelers as well as farmers and merchants. "Queen City of the Mountains" the burgeoning trading center and fashionable resort area was called. According to one Hall County historian, James Dorsey, as early as 1826 a traveler escaping the low-country heat of Savannah wrote, "This is as healthy a spot as can be found in any country."

Access by rail induced industry to locate in Gainesville. Factories, including the huge Pacolet

Manufacturing Company's New Holland Mill, built in 1900, and many smaller cotton mills, dominated the industrial growth that began in the 1870's. Later, the railroad aided entrepreneurs such as Jesse Jewell, who built his chicken empire from Gainesville beginning in 1936. In the years that followed, Gainesville established its claim as chicken capital of the world.

"Jesse Jewell went national from here," said former Georgia Mountains Museum Director Jim "Bimbo" Brewer, "the late forties and fifties were his heyday."

Poultry pioneer Jewell later gave his name to the unofficial dividing line between north and south Gainesville – the line that still defines white and black residential areas – the Jesse Jewell Parkway. Industrial expansion began again in the 1950's, and many of the new plants were sited on the south side, encircling New Town.

As Gainesville's industry was set to expand afresh, the New Town Florist Club was being born. It was 1950, and the homes along Desota Street were being purchased by the residents – for real this time. Starting the Club was Jack Ware's idea, recalls Mozetta Whelchel. A neighbor had been collecting money door-to-door to buy flowers for funerals. But sometimes the collections came up short. "So Ware told his wife," Whelchel explains, "'Why don't you-all start a club, a flower club,' and that's where the New Town Florist Club started from."

Annie Lou Ware, Cantrell, Whelchel, Elzora Davis, Laura Barnett, Colean Castleberry, Cecil Cleveland, Maggie Johnson, Tommie Phillips, Dora Harbin, Charlotte Garner, Ruby Wilkins and Amanda Keith were all founding members. The dues at that time: "Ten cents every meeting," Whelchel remembers. The women met in each other's homes, and held programs in various churches, willing to give aid where it was needed.

"It is beautiful the way we do"

"We started collecting money to buy flowers", says Keith, who is seventy-nine, "and then we started helping people when they was sick." At funerals, "we are flower bearers. We dress in white or black and have a beautiful red rose We get to the church first and line up and get the flowers and we march in behind the casket. ... We try to have at least six or seven flower bearers; sometimes it's more than that.

I think it is beautiful the way we do. When we come out of the church, they give us the flowers and we bring them out and we hold them. The Club members get in one or two cars together. We ride behind the hearse. The policeman's in front, we're behind them or behind the pallbearers. We like to stand and for the family to see us and we are beautiful."

"It makes you feel good. It did me when my husband died, and I hope they do the same thing for me. After the funeral 'cession and everything, everybody goes in different directions, and if somebody doesn't have a car, we make sure they get to their destination. ... Usually somebody goes to the house and makes sure the family is fed and sends them a little taste of change."

Together, Club members helped each other and

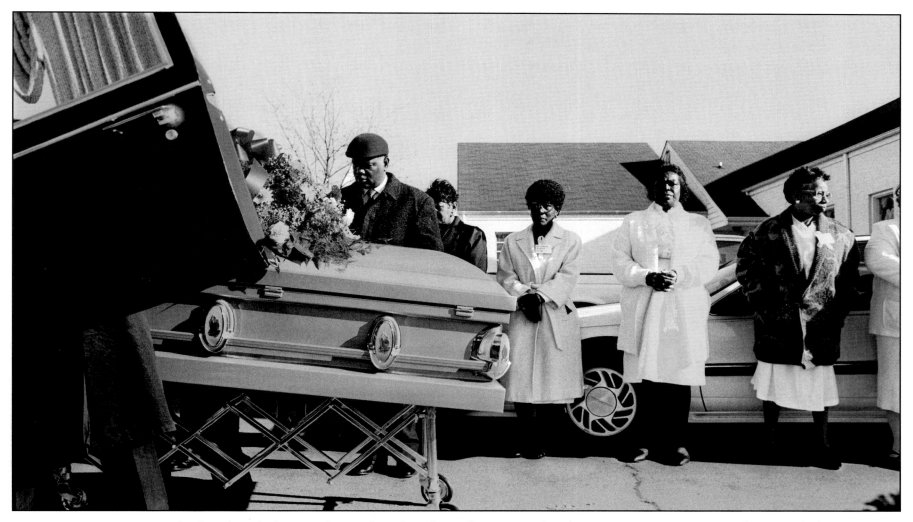

Rev. Eugene Green helps load the casket of Roland Waller into the hearse as Newtown Florist Club members Faye Bush, Mae Catherine Wilmont, Florence Tanner and Eloise Price look on.

Anna King sits on the porch glider of her grandmother Charlotte Garner's Desota Street home.

their children weather the difficult, hope-filled years during desegregation. When the young people were still denied the chance to take part in after school plays and sports at the newly integrated school, the Club began forming youth groups. One group of girls, who chose the name BOSSETTES (Beautiful Original Soulful Sexy Everlasting Tough Together Sisters), put on plays. The Ebonettes, echoing Amanda Keith's comment and the Movement's theme, "we are beautiful," started the Miss Keisha beauty pageant.

In a symbol of the larger community's growing recognition of their leadership, Club members were pressed into service as deputies to help restore calm in New Town during a 1972 riot following the stabbing of a black teen by a thirty-year-old white man at a high school football game in Gainesville City Park. "The scars ran forever and ever deep," says Brewer, a white Gainesville native, reflecting on the simmering tension that fall.

"They put badges on 'Miss Ruby' Wilkins and myself and several more of us," Bush explains, "and we went through the streets to try to tell the kids that things was going to be better and we were going to see that they did something about it."

Leaders within the community emerged. The most revered among them was Wilkins, a guiding force until her death in 1989. Under her caring hand, children were nurtured and youthful problems listened to. "Her yard became the south side recreation center," said Rose Johnson, who played there as a child. Wilkins got out and played ball with the kids. And she carried the young people right along with her to City Hall to protest the lack of recreation facilities for black youth. With her leadership, after more than a decade of insistence, and with much of the hard physical work of clearing the land done by the women themselves, the "raggedy basketball goal" in her side yard was replaced by a community park. Wilkins's philosophy: "If you want things to change, do something about it. If not, shut up."

"I can remember several people that was running for office would come to my mother's house and sit down at her table and eat. They would call on her," says Barbara Wilkins Jones, Wilkins' youngest daughter, who was born in the house on Desota Street. "It was mostly white [politicians], at that time, it was more white because no blacks" were running,

Speaking out was not easy. "They got threats from the Klan, with the park, with the Dr. King march," which the Club organized in Gainesville beginning in 1970, says Jones. "My mother, she walked with Dr. King, she marched in Alabama, she went to Washington, she was in the NAACP and in SCLC, so she could see the struggle all along."

Spurred on by Ruby Wilkins' fighting spirit, and in concert with the civil rights movement, the Newtown Florist Club began to take on other issues. During the years of urban renewal, the Club fought to get sidewalks and paved streets, to eliminate poor housing and outhouses – and to halt toxic pollution.

"The stench was intolerable"

The completion of Buford Dam on the Chattahoochee River in 1957 had brought a ready source of power and water to Gainesville, further driving industrial expansion. Fifteen new industries located in Hall County during the 1950s and twelve companies expanded during the decade, according to a 1962-era labor market report. The "Gainesville Survey," prepared to attract industry to the area, promised a workforce of "almost entirely Anglo-Saxon origin" and "not generally sympathetic to the union movement."

Planners anticipated other corporate concerns. "For a plant having a large effluent disposal," the report states, "approval of discharge into streams is required by the Georgia Health Department and other State Agencies. State laws on this subject are flexible and each case is judged individually on basis of circumstances involved. Generally, policies are very liberal and favorable to industry."

The first formal protest against pollution by Club members was filed with the aid of Georgia Legal Services against Purina Mills in the mid-1970s. Since its construction in 1954, the Gainesville division of the nation's largest animal feed producer had prompted neighbors to complain of a yellow-brown dust that settled on cars and windows overnight. If children played outside for a long time, residents say, they would come in with their eyelashes and the tiny hairs on their arms furry with grain dust.

An open sewer ditch ran between the homes and the street. When it rained, "It would just flood underneath the houses," says Desota Street resident Jerry Castleberry. The stench near the plant was "intolerable."

On a dry, hot day in June 1975, "there was a flow of water in the catch basin which came in from under the railroad tracks from [the] direction of Purina Plant," though it had not rained in more than 48 hours,

reported city sanitation inspectors E. Evanoff and Bill Butler.

The production manager at Ralston Purina took the two inspectors on a tour of the back side of the Purina Plant. "At two places back of [the] plant, contaminated water was draining into [a] ditch then into storm drain," summarized Butler's report. "I checked the water for odors and found it to be repugnant." Decaying grain and fermenting feed dust fouled the water and the air.

The Environmental Protection Division (EPD) of the Georgia Department of Natural Resources found Ralston-Purina in violation of the Georgia Water Quality Control Act. Corrections were ordered.

But months later, reports show "the drainage from around the tallow and molasses tanks discharged to a drainage ditch to Limestone Creek." Releases of waste came from fish and fat tanks, surface runoff and fuel oil tanks. Water quality officials also noted the contamination of the Mill Street sewer. More than a year after the original complaint was investigated, the open ditch behind Ralston-Purina was filled in and storm drains were installed. Standing pools of sewage and industrial waste along the railroad tracks were drained.

But at an August 1977 public hearing on community development plans – more than two years after the original complaint – residents again reported waste feed washing down and fermenting in the storm drain pipes. Odors were especially strong at the catch basins on Desota and Mill streets. The neighbors persisted, and eventually pumps were installed at Purina that reduced the problem.

Faye Bush (right) stops to check on an ailing Ruth Cantrell at her Desota Street home.

Successful in its expanded role, by 1978, the New Town Florist Club incorporated. Members pressed the demand to reopen the former all-black E. E. Butler High School as a recreation center for young people. "I remember the day when (Parks and Recreation Director Bill) White brought the key to the Butler gym, fifteen of us walking out to Butler behind Miss Ruby," says Rose Johnson, one of the young people Wilkins was grooming to be leaders. "It was a victory of sorts."

There were other victories. After five years of hard work by the New Town Florist Club and the Lanier Centennial Women's Club, the Neighborhood Cultural Arts Center opened near the Harrison Square apartments in 1982. Each new challenge proved to be a valuable lesson for the leaders Wilkins was training. Johnson remembers sitting on Wilkins' front porch: "Every time, we would share frustrations about the lack of sensitivity or even concern" about the shortage of city resources for what was coming to be called Newtown. "It took ten years worth of asking" to get land next to Wilkins' home designated as a community park, Johnson says. "It took all of that to get what little that you see." But Wilkins instilled a tenacity that would serve the community well in the struggles to come.

"That put us on a wonder"

By the late 1980s, carrying out the original mission of caring for the sick put members of the Newtown Florist Club in a unique position to notice a high level of respiratory ailments and other illnesses

Mozetta Whelchel and her son Jerome.

and an extraordinary number of deaths due to cancer and lupus in the neighborhood.

"That put us on a wonder," says Mozetta Whelchel. "I've lived here for forty-three years and can tell the story. I have lost all my family but one son, who's thirty-eight years old and working hard." Whelchel lost two of her children – Moselee at age sixteen and Deotris at age twenty – to lupus, a chronic disease that attacks the body's immune system.

"I first lost my daughter, and then, I thought, well, it's over, you know. And then about two to three years later, my son come down with it. I just had a hard time. It was hard for me to go through it, and then it come down to the time that I had a tumor under my

Never far from an industrial site, Newtown residents play ball at Ruby Wilkins Park.

brain. Like to died. Well, I guess He had something else for me to do. And then come down my husband had cancer. I just had a rough road to travel."

"About money, all I made went to the hospital," she explains standing next to her home just across the railroad tracks from Purina. "You have to be strong, you've got to let God...," her voice is swallowed by the long deep daily moan of the train horn.

Faye Bush, the Club's president and Whelchel's younger sister, at sixty-two, describes her medical problems over the decades: a baby delivered stillborn, a cyst on her vocal cords and a triple heart bypass operation. She continues to suffer with lupus of the skin. She and her children nursed her brother, Ace Johnson, until he died of throat cancer in the 1970s. He was in his middle forties. "My brother died, he died here in the house. He had got so small we could pick him up like a baby," she says. The family tried to get him into a cancer hospital in Atlanta. "About three months after he died, they called and told us they had an opening."

Booker T. Harbin

Their older sister, Dora Harbin, lost her husband, Booker T., who died of colon cancer March 1, 1988. He was eighty. A year and a half later, their son, William, nicknamed "Dub," suffered a heart attack at age fifty-nine after several years of declining health, and died September 27, 1989.

Whelchel and Bush live down Desota Street from each other in Newtown. In 1990, young neighborhood residents Marsha Browner and Selena Stringer conducted an informal survey. They found that, among the seventy-five residences in Newtown, eighteen neighbors in the forty homes reached have contracted cancer, three people have been diagnosed with lupus, and twelve have asthmatic bronchitis. Others harbor aching memories of stillbirths and miscarriages.

"There seems a shadow on the day"

Ruth Cantrell lost her sister, Lloyd Singleton Smith, at 62, to lung cancer in February 1980. "There seems a shadow on the day," mourners read from a poem at Smith's funeral. Cantrell's daughter died – of breast cancer – at age thirty-four, in California. Cantrell battled cancer herself from 1980 until her death in January 1995.

Further down Desota Street lives Wilkins's granddaughter, Pauletta Wilkins, who the neighbors call Missy. Her mother, Lonnie Ann Reed, died at age fifty-six, in July 1991, of lung cancer.

Pat Moorehead is another Desota Street neighbor. "I've had a couple of miscarriages, and I mean there was no reason as to why," she says. "My mom – she didn't think about it until later – she had TB. She was raised on Cloverdale. My dad died from cancer."

Knowledge about the illnesses rife in the several-block area was limited.

When Jerry Castleberry got sick, "They asked

Rose Johnson and Newtown Florist Club members dedicate a monument to residents who have died.

Jackie Mize takes a hair sample from Betty Ann Lewis so it can be tested for levels of heavy metals.

me if anybody in my family had lupus and I wasn't sure. My mom was always sick. I remember her being sick since I was five years old. I didn't even realize until after, well, long after she died, in fact it was last year that I – Faye would always tell me that I had lupus. I told her she was crazy."

Lupus caused Colean Castleberry's death February 2, 1981, explains her son. "I remember her flopping down in the chair, and she could hardly walk sometimes, and now I'm going through the same thing, so I can relate to it. Fatigue. Loss of hair. Some depression. Just the constant pain. Never had real good days."

The symptoms he describes are typical of those with systemic lupus. The immune-system disorder affects more than 500,000 Americans and is associated with arthritis-like chronic inflammation of the joints. More serious complications involving internal body

Colean Castleberry

organs – kidney, heart, and brain – can result. Another type, lupus of the skin, or discoid lupus, produces skin rashes and irritation. Lupus is nine times more common among women than men and occurs more frequently among black Americans than among any other racial group in the country. The disease occurs at a higher rate in Georgia and the Southeast than in other sections of the country, says Medical College of Georgia Chief of Medicine Dr. John A. Hardin.

The cause of lupus remains a mystery to medical researchers. "There are very few real facts," says Hardin. "There are a number of environmental-like factors that can affect lupus." For example, the "drugs, hydralazine (used to treat high blood pressure) and procainamide (used to control cardiac arrhythmia) are recognized as inducing a lupus-like illness in a small fraction of patients who receive them, but most people who get those drugs do not get lupus." Some studies have associated lupus with the inheritance of an "autoimmune" gene and female hormones. One study hypothesized that excessive exposure to an amino acid in alfalfa sprouts was a cause, but the results have not been replicated. "What is lacking," says Hardin, "is really good epidemiology studies and really good information about what kind of environmental factors might be out there."

Some would doubt an environmental link in Castleberry's illness, because lupus can be hereditary, and his mother lived her youthful years elsewhere. But several of the people with lupus in Newtown had no family history of the disease. Most had never heard of the disease until Whelchel's children got it in the 1960's.

Mae Catherine Wilmont, who grew up down the street and around the corner on Elm Street from Bush and Whelchel, first began wondering when she was diagnosed with lupus in the fall of 1991. "Because nobody in my family has ever had lupus," she was puzzled. "I have never been sick. I didn't have any

Pauletta Wilkins holds a photo of her mother, Lonnie Ann Reed, who died of lung cancer in July, 1991.

illnesses you would think that would be related to lupus-like malaise, tired all the time."

Wilmont is a licensed practical nurse. So, after her doctor said she had lupus, she read up on the disease. "I've got two types. I got systemic and I've got discoid, which is lupus of the skin. I just figure it's got to be the environment."

The illness forced her to cut back to four hours a day on her job in the mother and baby unit at a nearby hospital. When she first returned to work in May after her illness, the hospital sent her home on disability. Then, back on "light duty" work, she says, "They had this lady to follow me. She wrote me up for everything I did. I got four variances

Mae Catherine Wilmont

in a week's time. And I've been there for 13 years and never had a variance. I always had good evaluations."

In August 1993, the hospital put Wilmont on an involuntary leave of absence. Three months later, she was terminated. She expected more understanding from the hospital "... for the simple reason they are in the medical field. They should know about these kind of things."

Like others, Wilmont feels sure the environment is to blame, but as community after community around the nation has found, links between exposure to toxic

substances and disease are agonizingly hard to prove. When the informal 1990 self-survey revealed the alarming numbers, residents pressed hard for a health survey by the state.

The state health survey, conducted by Georgia Department of Human Resources (DHR) epidemiologists Tom McKinley and Dr. David Williams in 1990, found unexpectedly high levels of mouth and throat cancer (four cases where 0.9 would be expected). Residents were outraged at the state's unsupported conclusion that "lifestyle" – smoking and drinking by Newtown residents – was the cause. At least two of the throat cancer patients did not smoke or drink.

Interviewed four years later, McKinley stood by his conclusion, but confessed "we didn't try to quantify" alcohol use. There was "not much quantification" of tobacco use either, he says, though he admitted such data is necessary to link smoking habits to cancer. Were the environmental links explored? One cannot go to Newtown without being struck by the encroachment of industry; but the interviews were carried out by phone. According to McKinley, the team did talk to EPD about the toxics and dust samples. Though correctly noting that evaluating environmental exposure is "extremely difficult to do," McKinley added, incredibly, for an environmental investigator, "You have to look at what you can see."

Residents sent the state's study and findings to the Citizens Clearinghouse for Hazardous Wastes (CCHW) and the Legal Environmental Assistance Foundation (LEAF). Stephen U. Lester, science director for the CCHW, analyzed further problems with the study's methodology. Lester criticized the DHR survey for relying mainly on answers provided by phone during a three-week period, failing to separate people by how long they had lived in the community, and using inappropriate comparison groups and sample sizes. Asked about scientific sampling techniques in selecting residents to study, McKinley responds that his team conducted fifteen- to twenty-minute interviews with "everyone we could find."

Angry that documented high cancer rates were officially dismissed with a slur against "lifestyle," Club members pushed for independent health assessments.

"A double jeopardy"

Newtown's compelling story and the Club's diligent recruitment efforts succeeded in bringing a number of health and other professionals to volunteer their services in an effort to pinpoint health problems and explore environmental links. Dr. Elizabeth Bowen, former assistant professor of family medicine at Morehouse School of Medicine and past national president of Physicians for Social Responsibility, volunteered to screen close to fifty individuals in Newtown.

"In the black community there is a double jeopardy" from toxins, says Bowen. "Not only are African Americans exposed to hazardous materials excessively," she says, but other health care issues may leave the population more vulnerable. For example, less access to early diagnosis, lower levels of insurance coverage, and high costs of treatment can compound the effects of greater exposures to toxins.

Bowen looks beyond conventional diagnostic techniques, testing hair, blood, and tissue samples for traces of biological levels of exposure. She identified high levels of metals in hair samples of six or seven residents, too small a number to be statistically significant. Bowen acknowledges her methods "haven't been endorsed by the mainstream medical community," but feels that clinical ecologists need to employ new techniques. To identify a source requires painstaking effort, she explains. Air conditioner filters, vacuum cleaner dust bags and work clothes may need to be tested; perhaps the family member who does the laundry is exposed to chemicals brought home on work clothes. "It takes a lot of detective work," says Bowen. "It is one thing to show an association, to have a child with lead intoxication and learning difficulties. It is another matter to show cause and effect, to remove the lead and demonstrate that the child is then able to pay attention."

Her main theme: to effectively treat disease, the source of environmental toxins must be eliminated.

Another health survey of Newtown was designed by Dr. Frank Bove, an epidemiologist who works for the federal government. Trained volunteers, including Newtown residents and students from Vanderbilt University and Clark Atlanta University, conducted and tabulated the results, identifying cases of lupus not found in prior studies.

To further understand the high number of lupus cases, Club members sought the help of Emory University Environmental and Occupational Health specialist Dr. Howard Frumkin. Frumkin and medical student Tarik Kardestuncer looked at how many people in

Keosha Holcomb rests from swinging at Ruby Wilkins Park, which is next to an industrial site.

Garrick Tywee Hendrix, Ty Hood, and Deotris Mize represent the future generations of Newtown.

Newtown have the disease and the rate new cases are developing. They found a six-fold increase over highest available comparisons in the proportion of Newtown residents with the disease and a rate of occurrence of new cases nine times greater than the highest reported rates for African Americans. While forthright about several limitations of the study, Frumkin also cites conservative assumptions that were made, which make it likely their conclusions are understated. Researchers concluded that the "findings in Newtown suggest that long-standing exposure to industrial emissions is associated with an increased risk of being diagnosed with lupus."

Their report recommended further disease research, environmental sampling of both air and water, local health providers becoming more familiar with environmental health issues, and medical screening to detect both evidence of exposure and early disease.

Part of what baffles residents and health professionals – in addition to the documented higher lupus and mouth and throat cancer rates in Newtown – are reports of such a variety of different kinds of cancers and other illnesses, including bronchial ailments. "There are almost no exposures known that can cause so many different kinds of diseases," says Frumkin.

Very little is known about exposure to multiple toxins. There may be a "synergistic effect," which occurs when chemicals act together to have a greater effect together than they do individually, Frumkin explains. "Nobody really understands very well the effects of mixtures, but we have to assume that they are at least as bad as the effects of individual chemicals when added together, and maybe worse."

Health studies usually do not provide the kind of certainty researchers seek or the community needs to prove an environmental link. Studies "often aren't very definitive," Frumkin continues. "At the end of the day, you are left with your suspicion and your concerns. Unfortunately, that is the state of the art."

But even without definitive proof, victories are possible. Even if the studies do not reveal what is making residents sick, "it's completely fair to limit your level of exposure," he emphasizes. "The victories are often much more on the level of communities getting together, taking charge, organizing."

And that's where the Newtown Florist Club excels, getting together, taking charge, and organizing.

"If you want something to change..."

In 1990, residents focused in on the closest, noisiest and most visible offender – the junkyard – whose owners were seeking a permit to expand. From its entrance on Athens Highway the junkyard stretches along the railroad tracks behind a dozen homes in Newtown.

"Sometimes during the day when the temperature is high, it's hot... there's an odor," says Moorehead, whose back yard abuts Gainesville Scrap Iron and Metal. "Some days, we can sit out on the back porch and see mice, rats jumping, the weeds and hedges had grown up over the fence and it cause... an overflow of water into the backyard... the sight... the scenery, it's awful." Several fires have been reported at the waste and scrap site, including an explosion on April 26, 1993.

Trying to prevent the expansion, in January 1990, Christine Young, Louise Hudson, Jonathan Butts, John D. Wilkins, Clyde E. Wilkins and Tracy Lamar Jones sued junkyard owners Harrison Haynes and Sanford Loef, the Seaboard Railroad Co. and the City of Gainesville. But their request for a temporary restraining order to stop the junkyard's growth was denied in April 1990 by Hall County Superior Court Judge Rick Story.

Complaints continue, but scrap yard owners are shielded by the "60 percent rule," a regulation that allows scrap businesses to be "self-monitored" if 60 percent of incoming scrap metal is recycled every 90 days. Only if such businesses are reported will EPA investigate, asking for manifests and records. But, watching from porches and back yards, homeowners have to endure the growling backhoes shoving around the piles of metal scraps, some of which they say remain long beyond the time limit. Visible through the wavy screen of the thin plastic mesh "buffer fence" that the junkyard was required to erect are a disabled cement truck, dead refrigerators and cars, rusted boilers and an ancient tractor-trailer cab. The whirring blades of an old industrial fan glint in the afternoon sun.

"When the city of Gainesville allowed the junkyard dealer to expand behind our homes, Newtown Florist Club marched its big march," says Rose Johnson. "We protested the junkyard dealer, and I remember the city manager being over in the community all day that day because he was getting pressure from downtown about this march."

"Shortly after that march," Johnson continues, "the city revised that parade ordinance to restrict all of the marches to around the square in downtown Gainesville." Johnson continues, "And it was also shortly after that that the Klan emerged and requested a permit to march through the black community, to 'rid our community of drug-infested crime.' And the City of Gainesville issued them a permit to do so."

"You can't bring them hoods to our front door"

"At that point, there were two projects going on in the community," explains Johnson. "One was the campaign against the junkyard, and the other was the first anti-drug campaign that we had ever initiated. But because the Klan emerged and the city issued them a permit to march through the community, we had to stop dealing with drugs and contamination and deal with the Klan, and I think we spent six months or more."

The Klan had resurfaced in north Georgia during the 1980s. In January 1987, in neighboring Forsyth County, a white mob threw bricks and bottles at Martin Luther King Day marchers. The mob action prompted a national outpouring of concern, including a march that brought more than 25,000 people one week later. Newtown Florist Club members worked together with white residents to develop a constructive response to the threat of Klan violence in Gainesville.

Bush remembers the Klan coming through when she first moved to Newtown, around 1948. "I was

Newtown resident Marcia Browner, who formed Black Teens United for a Future (BTUFF), volunteers to help the Newtown Florist Club.

Debris fills the Gainesville Scrap Iron and Metal lot adjacent to the Newtown community.

staying with Mozelle at that time, before my mother and them moved here. They came through here and they had sheets on their head and I was so afraid. I never seen anything like that," she says.

Neighbors say the Klan stayed out of sight around Newtown for two decades after the 1960s, when a hooded gang protesting the opening of Arthur Lipscomb's Athens Highway business – the first black-owned service station – was dealt a surprise. In a turn-about of their own tactics, "a cross was burned on them," Bush says. And after neighborhood youths rocked invading cars, she continues, "They didn't come back. That's the last time I knowed them to ride through the neighborhood."

But in 1986, two klansmen, including "Great Titan" Daniel Carver, of the Invisible Empire, Knights of the Ku Klux Klan, who lived less than ten miles away in Oakwood, set up a demonstration on Sycamore Street, near Newtown.

Jackie Mize articulates the Club's position on the Klan marching through Newtown. "March uptown all you want to, we won't even go up there ... But you can't bring them hoods to our front door. You can't do that." Mize continues, "The law says if you are going to incite a riot, you cannot march.... That's what the Ku Klux Klan is doing, inciting a riot. They cannot march just simply on that basis."

Each year since 1970, the Newtown Florist Club has sponsored an annual Martin Luther King Holiday March and Observance. In 1991, the Gainesville City Council enacted a parade ordinance, which would have prevented the Club from marching on the traditional

An effigy hangs from a tree over the yard of a alleged Klan member near Newtown.

During the "Toxic Tour" in 1993, Faye Bush places a bow made of black ribbon on the home of a Newtown resident with an illness.

King Day parade route from the downtown square through the black community. Mize and others were outraged. "Then you gonna deny what we been marching twenty years for Martin Luther King ... you gonna deny us our right to march because the Ku Klux Klan couldn't march over here."

So, with aid from the Atlanta-based Center for Democratic Renewal and lawyer Georgia Lord, the Club sued and won in court – preventing the Klan from marching through the black community and retaining the right to use the traditional King march route.

Victorious on the parade ordinance, but diverted from the junkyard fight, Newtown residents lost their battle to keep Gainesville Scrap Iron and Metal from expanding behind their back yards. Attention now returned to the campaign against toxics.

After years of watching sinkholes develop in their back yards and finding trash when digging in their yards or basements, residents also sought confirmation that much of Newtown was built atop a landfill – the pre-1936 city dump.

Not until October 7, 1993, was that landfill's existence confirmed by state officials. "The vast majority of the 'landfill' is between Mill Street and Cloverdale and from east of Elm to Harvey Street," along a tributary of Limestone Creek, reported the state EPD. "The landfill appears to have been an open 'dump' with burning."

Samplings from a methane survey, twenty-one soil borings and two groundwater monitoring wells revealed the anti-termite chemical chlordane and "a slightly elevated lead and chromium content." But "the elevated levels are most probably due to natural causes, because the area is within a mineralized zone," the report concluded.

"We found no evidence that the 1936 landfill poses any health or environmental threat to the residents of Newtown," wrote Environmental Protection Division Director Harold F. Reheis in the 1993 report.

The quantities of lead, chromium and zinc noted in the EPD data are not easy to interpret. "They are higher than you would expect," says University of Georgia geologist Dr. John Dowd, "but they may be naturally occurring." The EPD test did not distinguish between Chromium VI, which is harmful, and Chromium III, which is a necessary nutrient. Also, notes Dowd, the soil samples that were tested for metals were taken in Bush's and Whelchel's yards on Desota Street, not on Cloverdale and Mill Streets where the old landfill was located, and no comparison tests to establish background levels were made.

A growing number of technical and legal experts volunteered their time to assist residents. Dr. Ed Mayhew, a biologist at Gainesville College, showed young people in Newtown how to sample water in nearby streams and at Lake Lanier. Dowd and his advanced hydrogeology students sampled soil and groundwater in Newtown, gathering data to help residents independently assess the hazards. The samples gathered revealed little evidence of metals from the old dump that would pose a health hazard, but raised new questions. The students did note high levels of lead under one yard on Cloverdale Street and an unusual bleached area of soil near the Southern Railroad tracks, where the red Georgia clay had mysteriously turned white.

"The damage had been done"

Some residents feel they have been getting a double dose – living in Newtown and working in the nearby factories. Roland Waller worked most of his life at Leece-Neville, a starter motor factory, which he could see from his back door.

"It was in 1969, May the 13th," that Waller began work there. "I got sick on June the 2nd, 1992," he said. "I was over there twenty-three years." He worked over washers, inhaling fumes from ten steaming 300-gallon vats, with "acid, pure acid, trichloroethane, fumes. Had another one called phosphor. ... They banned them in '88," he said. Waller was a steady participant in efforts to stop toxics in Newtown before he died – of congestive heart failure – on January 29, 1994, at age 55.

Waller worked in what he described as one of the most toxic areas of the plant, the Butler Building, which workers called the "hot box." The heating element under the vats brought air temperatures to upwards of 105 or 110 degrees in the sweltering summer months, he remembered. "We didn't have no protective gear until Prestolite took over. See, we wasn't thinking about that at that time. I wasn't thinking about it. Nobody else wasn't thinking about it. That was up to the supervisors and the management to think about."

The chemicals that swirled around the workers were not confined to the plant. Prestolite, which operated the Leece-Neville Plant until 1992, reported releases of 9,934 pounds for the year 1990 of xylene and combined releases of 4.38 tons of 1,1,1-trichloroethane and freon in the process of making AC motors, diesel starter motors and regulators for alternators. Chronic exposure to xylene, a flammable solvent, may damage the liver, kidneys, skin, eyes and bone marrow. Trichloroethane, also a cleaning solvent, can cause mutations in cells, and damage the liver, kidneys, and skin. Acute exposures can cause dizziness, loss of consciousness, irregular heartbeat, and death.

Roland Waller, who died in 1994.

"Primary emissions are VOCs (volatile organic compounds) from paint booths and varnish dip tank and baking ovens," reported investigator William C. Montgomery to Ron Methier of the state Environmental Protection Division (EPD) about the Leece-Neville site in 1989. Chemicals were used in soldering, welding,

Mae Catherine Wilmont with grandchildren Deshayla Bush and Shundria Bush at her Newtown home.

Environmental activist Sheila O'Farrell (left), Faye Bush, and geologist Ray Preston do soil tests in the backyard of a Newtown home.

paint-washing, paint booths, baking ovens and the varnish dip tanks.

Leece-Neville maintenance supervisor John D. Harrison certified to the Environmental Protection Agency as early as 1982 that all wastes were placed in drums and shipped off-site, originally to the Hall County Landfill and later to Chemical Waste Management (Chem Waste) in Emelle, Alabama. The plant was found by the Georgia EPD to be in compliance in 1984. "Recommend approval of the permit to operate," wrote EPD staff again in 1989.

But Waller strongly believed the plant was unsafe. In an interview six months before he died, Waller described his work at the plant. "You had to stand over these vats in order to get the parts cleaned out," he explained. "You didn't know what, exactly how it would affect you in the long run." Waller said fellow workers told him, "before I got sick, 'keep working 'round those vats, it could cause a stroke or either a heart attack.' But I didn't realize that." Waller had been diagnosed with tuberculosis as a child and spent time in Valley State Hospital in Rome, Georgia, but had been released and had been healthy since moving from Eatonton in Putnam County to Gainesville in 1958.

"So you breathing these fumes in a length of time, it's gon' take effect on your heart muscle and your lung. And that's what I think my condition come through. ... I can't explain it, but you could take a piece of aluminum, just something aluminum and throw it over in there, you wouldn't see it no more. ... You turn your back 'bout thirty minutes, you go back you ain't got no parts, 'cause it done fell down in the acid. It ate that wire up."

"The chemicals was so hazardous 'til they took and sealed all drains so if someone spilled on the floor or something, it would not go in the drain," Waller said. "After this other company took over, they dismantled those tanks, but the damage had been done."

The Leece-Neville site is a pre-remedial Superfund site, subject to the 1980 environmental law requiring investigation and clean-up of toxic sites. First recognized as a hazardous site in 1980, the now-closed plant lies roughly fifteen hundred feet from the back yards on Desota Street.

Waller counted off names of a dozen former fellow workers with health problems. Among them was Mozetta Whelchel's husband, Lee, who worked at Leece-Neville for seventeen years. Before he died of lung cancer at age seventy in 1990, he urged others to join the emerging efforts to rid the community of toxics.

"An industrial fallout zone"

Though connections between exposure to toxics and illnesses or deaths are especially difficult to prove, the unequal siting of facilities is obvious. "The toxic burden is on the black community. That has to stop," wrote ECO-Action, an Atlanta-based group that provided strategic assistance to the Newtown community in the early 1990s. ECO-Action helped Club members research the toxic emissions coming from surrounding industry. Together, they developed a toxic profile of the community using the toxic release inventory (TRI), a database published annually by the state EPD as required by federal law, compiled from industry self-reports of

Amanda Keith with great-granddaughter
Victoria Amanda Thomas.

releases and transfers of chemicals to the environment.

The TRI has shortcomings as a comprehensive measure. The Office of Technology Assessment (OTA), the scientific research arm of Congress, "estimates that due to the relatively short list of chemicals and industries covered, the TRI reports on roughly five percent of toxic pollution," writes the National Campaign for Pesticide Policy Reform. Nor does the TRI record actual human exposures.

Simply being listed on the TRI implies no violation of emission standards. But it does mean that the company emits or stores chemicals that are known to cause significant adverse acute health effects, including cancer, serious or irreversible reproductive dysfunction, neurological disorders, genetic mutations, or have significant adverse effects on the environment.

By using TRI data to carefully map the companies reporting emissions, members found that thirteen of the sixteen industries in Gainesville that reported toxic releases in 1990 were located south of the racial divide at Jesse Jewell Parkway. What's more, all of the hazardous waste generators that year, companies that produce waste as a by-product that they are generally disposing off-site, are located on the south side.

"The toxic profile documented the disproportionate burden, and ended speculation regarding the extent of the racial inequities," says ECO-Action director Carol Williams. "The toxic profile map gave a visual tool to illustrate what residents knew: the south side was getting dumped on."

Using the toxic profile, Newtown residents were able to press environmental authorities for more facts about potential exposures. Their correspondence with state regulators revealed additional hazards.

Piedmont Labs, on Old Candler Road, which makes cosmetics, soap and other detergents, and is a "large quantity generator of hazardous waste," according to EPD, reported releases of trichloroethane, ammonia, ethylene glycol, glycol ethers, methanol, toluene and xylene, totalling 165,631 pounds in 1992. On January 28, 1992, a small fish kill reported in Flat Creek was traced by the state EPD to a white milky material which led to the back door of the Piedmont Labs packaging building. The EPD did not pursue enforcement

Today's Newtown Florist Club members includes (from left) Eloise Price, Florence Tanner, Geraldine Collins, Mae Catherine Wilmont, Sara Nash, Faye Bush, Mozetta Whelchel, Annette Westbrook, Jackie Mize, Coretta Young King, Dorothy Kea, Pauletta Wilkins, and Peggy Brown.

action although they found that an employee of Piedmont Labs was responsible, for "taking a short cut," reported EPD Director Harold Reheis in a letter to the Newtown Florist Club.

A later spill was reported, in mid-February, 1995 that resulted in a second Flat Creek fish kill. According to The [Gainesville] Times, pesticides were found in state soil samples. This time, under a consent order with EPD, Piedmont Labs agreed to restock the creek and pay investigation costs, but admit no guilt.

An industrial site does not have to appear on the TRI list to have been cited for breaking environmental laws. One such plant, Georgia Chair Company, was found in violation of the Georgia Rules for Air Quality Control in the operation of its wood waste-fed 1947-era boilers. State Environmental Protection Division investigators found the wood furniture manufacturer out of compliance with the 40 percent opacity limit. (Opacity measures smoke density; 100 percent is black; zero percent is clear.) The rest of the plant was found in compliance, but periodic problems with the fuel feeding system resulted in intermittent bursts of smoke, with opacity levels reaching 100 percent. "It's just like coal," says Bush, when industrial smoke clouds the sky.

Members also learned that the old Hall County landfill, a different landfill than the pre-1936 dump, a couple of miles from Newtown at Highway 129 and Monroe Drive, is a medium priority pre-remedial Superfund site. The landfill was identified as a potential hazardous waste site in 1982, and closed in 1985. The 161-acre site received "garbage, rubbish, ashes, dead animals, abandoned vehicles, industrial waste and special waste, such as tires, appliances and television picture tubes." For a time, the landfill accepted grease from the poultry industry and septic tank pumpings, as well as hazardous waste from Leece-Neville. Surface run-off drains into streams that feed into Allen Creek. Homes in Newtown receive city water, supplied by the Gainesville Water Works, which draws from the Chattahoochee River via nearby Lake Lanier. But the Club's efforts to uncover the dangers Newtown residents are facing revealed other families at risk as well. According to a site inspection report prepared in 1991 for the federal Environmental Protection Agency, sixty-seven homes in Hall County draw drinking water from wells near the old landfill, with the closest well just 1,400 feet away.

The TRI roster includes three chicken processing giants – Fieldale, ConAgra Broiler and Continental Grain (Wayne Poultry) – a smelly presence a mile-and-a-half southwest of Newtown. Poultry processing is now a $126 million industry in Hall County, and it continues to grow. Fieldale expanded in 1994; the company reports ammonia releases. ConAgra has reported releases of ammonia, arsenic, copper, manganese and zinc. Continental Grain has reported releases of copper, manganese and zinc.

According to industry's own reports, Hall County was in the top twenty of Georgia's 159 counties in reported toxic releases in 1990. The total releases reported in Gainesville in 1990 reached 587,664 pounds; nearly 75 percent of emissions (440,160 pounds) came from industries on the south side. The total waste generated, including wastes hauled off site, more than tripled between 1990 and 1991 from 587,664 to 1,851,445 pounds per year, according to the annual roster of reported pollutants.

More frustrating, no one knows the effects of combined exposure over a long period of time. At least six of the "dirty dozen", the worst pollutants identified nearly two decades ago by Mechanics Illustrated: arsenic, benzene derivatives, chlordane, chromium, lead and manganese, are emitted from south side industries. Even subtracting for wastes hauled off site, "at least 1.8 million pounds of pollutants have been released by nearby industries between 1986 and 1991 alone, federal reports show," reported The (Gainesville) Times in 1993.

The Club's work to call attention to the problem prompted extensive coverage in the local press. "It's an industrial fallout zone," wrote a reporter for The [Gainesville] Times in 1990.

With the discriminatory siting so clearly documented, in the spring of 1994 Newtown residents filed a complaint with the Office of Environmental Justice EPA Region IV to challenge the unequal siting of toxic emitting facilities under Title VI of the 1964 Civil Rights Act. Title VI bars discrimination by race, color, or sex by federal agencies in issuing permits and distributing federal dollars. Communities seeking environmental justice have begun using this legal tool to challenge permitting and zoning practices that result in disproportionate pollution in communities of color. Title VI has limitations – residents must challenge an action by a recipient of federal funds, usually a state or local government, and the action taken, such as a permit issuance, must have taken place in the past six months. And, claimants must meet a tough burden, proving discriminatory intent. Only thirty-six Title VI complaints have been accepted by EPA nationally as of April, 1997 and no environmental discrimination case has resulted in the withholding of federal funds.

The strict requirements regarding action by a specific recipient of federal funds in the past six months meant that the Newtown Florist Club's complaint, citing long-standing and multiple exposures, was rejected. Such limits on remedies do not preclude the city of Gainesville or the state of Georgia from taking action to reduce the toxic burden. City or state action could mandate pollution prevention and prevent new sources from siting in south Gainesville. Seeking redress, the Club turned to local authorities, pressing City Council for a moratorium on siting new industrial facilities. But an effort in the fall of 1992 to get the City Council to enact a moratorium failed.

At Purina, just across the tracks from Desota Street, the primary emissions reported are manganese compounds, zinc, and various pesticides.

Across Athens Highway on West Ridge Road, within earshot of Newtown, Cargill Feed Mill refines soybeans and extracts edible vegetable oil, soybean meal, soybean hulls and crude soybean oil. Cargill, which boasts of being the "planet's biggest grain trader" and is one of the nation's largest food companies, built its Gainesville plant in 1966.

Cargill reports storing between 10,000 and 100,000 pounds of phosphoric acid and sulfuric acid on-site and has reported no releases, according to TRI data. But state emergency response information reports show that

Cargill had two hazardous waste spills of sulfuric acid used in the waste water treatment operation, one on October 30, 1990, and one on September 30, 1992. Both spills were reportedly contained on-site; one was neutralized and discharged to the municipal sanitary sewer with city approval.

Alarm rose in 1995, when additional spills forced more than one emergency evacuation near Newtown.

"We have a right to know"

"Wheel of Fortune" spun on the television at the Hailey home on Friday, May 19, 1995, when Hall County emergency officials banged on the door and told the family to evacuate down the street – away from noxious odors that would send Paul Hailey to the hospital for five days. Hailey was one of more than thirty people hospitalized. Most were treated and released, for severe nausea, skin irritation, headaches, burning eyes and throats, dizziness, shortness of breath and sleepiness.

As emergency vehicles and emergency personnel swarmed on Black Drive, where the Haileys live, and Cooley Drive, also near Newtown, local hospital personnel were told to prepare for hexane poisoning. Environmental reports on the spill do not specify which form of hexane was suspected, but the chemical n-hexane may be the most highly toxic member of one family of neurological toxins, according to Patty's Industrial Hygiene and Toxicology. The odorous chemical rates as a severe fire and explosion hazard, and is associated with nausea, eye and skin irritation, dizziness, drowsiness, numbness of limbs, and bronchial and intestinal irritation.

Initial news reports pointed to the Cargill Feed Mill's periodic cleaning process as the source of the release. Cargill denied any responsibility for the spill, but sent letters to some residents offering to pay medical bills. The May 24, 1995 letters from plant manager Catherine Hay read that though it is "not determined what caused" the odor, "Cargill has made arrangements to pay your medical expenses associated with this odor."

The Environmental Protection Division's report on the May 19, 1995 event is inconclusive, "The identity and origin of the material remains unknown." But, on the night of the evacuation, the EPD investigator advised against sampling, "since the material was no longer detectable." EPD's report says, "Ms. Hay clearly stated Cargill's position that they did not feel they released hazardous materials, but that they had likely released an odor that passed through the community." While not finding Cargill guilty, the EPD did issue two recommendations: 1) EPD and Cargill identify better procedures for performing annual cleaning to minimize community odors, and 2) Hall County consider forming a Local Emergency Planning Committee (LEPC) to involve the entire community in emergency planning.

Later, in early December 1996, a smaller number of residents from the Black Drive and Cooley drive area had to evacuate.

"Two times last year, we was taken out of our homes. We have never been told what it was," Paul Hailey told residents and supporters gathered at St. Paul United Methodist Church to hear about a health study in 1996. The spills galvanized a new group, Concerned Citizens Against Pollution, formed by black Southside

Christine Young can watch and hear Gainesville Scrap Iron and Metal from her bedroom window.

Pat Moorehead's Desota Street home backs up to Gainesville Scrap Iron and Metal.

residents from the Black Drive, Cooley Drive and Jordan Drive area.

The Newtown Florist Club, Concerned Citizens Against Pollution, and residents of Newtown and the Black and Cooley Drive area filed suit on December 10, 1997, with the aid of the Lawyers' Committee For Civil Rights Under Law, charging Cargill with a violation of the Comprehensive Environmental Response, Compensation, and Liability Act of 1980 (CERCLA) in releasing the hazardous air pollutant hexane. The suit also charges Cargill, Inc. with failing to report the release to the appropriate authorities and with violating the air quality permit issued by the State of Georgia.

Lack of emergency preparedness for such events worries residents. Jackie Mize's friend who works at one plant "says if they make one mistake, one mistake, it could wipe this whole community off the map. They haven't even had an escape route for us. If the train was to overturn and some toxic chemicals were on that train ... how they gon' evacuate this city, this town? They don't even have a plan."

Hall County has an emergency management office, which coordinates preparedness planning with the city. Companies are required by state law to report to the fire department any explosive or toxic chemicals stored on site. But much is left to industry response teams and voluntary compliance by plant managers. "They are supposed to register this material with the fire marshal," said one city official, but "that doesn't mean everyone is going to do it."

"The plants are very cooperative here, they tell us periodically what chemicals [they have] and where they are stored," said Willard Baxter, former Hall County director of emergency management. "We are grateful for their foresight." But, again, there is no independent verification.

The fire department prepares "pre-plans" on area buildings, but relies on "chemical people" at each company. "We have no hazmat [hazardous materials] team," said Gainesville's acting fire marshal, Tommy Clark, in 1994. "It's so expensive and hard to maintain."

The 1995 spills and the formation of Concerned Citizens gave momentum to efforts to form a Local Emergency Planning Committee (LEPC), which collects information on toxic sources and gives another arena to demand accountability. Georgia has been slow to establish local emergency planning, though it has been required since 1986 under the Community Right to Know amendment to the Superfund statute. The LEPC must include representation from industry and the community, as well as local government, the fire department, the emergency management office, the local bar association, media, and local hospitals.

Belinda Dickey, Paul and Beatrice Hailey's daughter, is vice chair of the LEPC, representing Concerned Citizens, and Faye Bush represents the Newtown Florist Club on the Committee. Dickey hopes that the LEPC will increase "community awareness about what's around as far as hazardous chemicals. A lot of people aren't aware that we have a right to know. I wasn't aware, but we can request that information from employers and companies," she says.

The LEPC is for all kinds of emergencies, not just chemical spills, explains Dickey, "like a tornado."

"It's not going to be any miracle thing as far as bringing the hammer down on anybody," says fellow LEPC member Bill Brooksher, a local environmentalist who represents the Hall County Environmental Quality Board on the committee. "The LEPC just gives another avenue to make their case."

Cargill plant manager Catherine Hay chairs the LEPC.

While Brooksher feels high level industry involvement is essential, he feels progress in the LEPC's first year has been slow. "It hasn't gotten to the point where they are searching out violators, non-reporters and taking action," says Brooksher. "The ultimate goal should be to eliminate the risk." Residents want not only to reduce risk of accident, but to eliminate the hazards.

"Another power that the LEPC has is to inspect all these facilities and negotiate" for a reduction in hazardous chemical storage, said Sheila O'Farrell, former consultant to the Center for Democratic Renewal, who has worked closely with the Club. "Then they could say, 'This is unsafe with this proximity to the neighborhood to keep this much. Will you agree not to store this much on site?'"

As part of the effort to measure and limit exposures, residents have pushed the state Environmental Protection Division to post air-monitoring equipment atop the Fair Street School in Gainesville in early 1997. First promised in 1994, the one-year air quality study is expected to segregate and identify volatile organic compounds, as well as heavy metals and semi-volatile compounds, says Rafael Ballagas, of the state EPD. Why was Fair Street School, an elementary school which serves all of Gainesville, but is away from the industrial sites, chosen? Ballagas' only explanation is that public buildings provide free access to electricity. He admits the preferred site would be E. E. Butler Center, which is closer to the factories, but unavailable until after the building is sold.

Though air monitoring is welcomed by residents, skepticism remains because the placement of the test equipment seems guaranteed to skew results. Trust in state and local action is low because so little has been done to remedy known problems. The EPD did make available a few stainless steel canisters for "grab samples" which allow area residents to collect air samples for testing when odors are strongest.

Notably, and perhaps due, in part, to vigilance of groups like the Newtown Florist Club and Concerned Citizens, Georgia's 1995 toxic release inventory

Citizens, Georgia's 1995 toxic release inventory (published in April, 1997) reports emissions in Hall County, and the state as a whole, have declined. But 60 percent of the decrease in Hall County emissions is due mostly to a change in production processes at one Flowery Branch plant, ten miles away. And several industries that earlier reported are absent from the list.

Useful as it has been in identifying sources and mapping the disparate burden on south Gainesville, members are mindful that the Inventory understates the toxic threat. Not only is the data collected through voluntary company reporting, there is no oversight from EPD or EPA. "We have no authority to do independent verification on company reports," says Bert Langley, emergency response program manager at EPD.

Some localities have made progress in tackling toxics through local ordinances restricting emissions, but Gainesville city officials have not acted. "If anything, I think the situation is worse, as far as companies being recruited in secret" with promises of tax benefits and lax environmental rules, says Lillian Hall, a white Hall County environmental advocate who has supported the Club's efforts to fight pollution.

Despite repeated efforts to get the Gainesville City Council to act, and even with one black member on the five-member body, no local ordinances limiting polluters have been passed. The failure of local government to respond on environmental and other issues led the Club to initiate the Hall County Voting Rights Task Force.

"It all come down to the vote"

"I think it all come down to the vote," says Pat Moorehead. "I really think it come down to who you put in office." The same understanding motivated Rose Johnson to run for the City Council in 1990.

That's what got Sara Nash involved with the Newtown Florist Club, "when Rose was running." Johnson grew up in Newtown, and remembers as a teenager having early morning chats on Wilkins' front porch about what to do about cleaning up the grain dust. After a dozen years away in Atlanta, she returned in 1989 and got re-involved with the community and the Newtown Florist Club.

Johnson ran in Ward 3, the only majority black voting district in the city, which includes Newtown. According to the 1990 census, Gainesville's population is 23.5% black and 68.8% white; the voting age population is 20.24% black. In Gainesville's at-large system each Council member must reside in and run from a particular district, but is elected citywide. This post-at-large system is a variation on at-large election schemes, which were widely implemented in southern communities in response to the passage of the 1965 Voting Rights Act. It allows the majority of white voters to consistently defeat the candidate supported by the black minority.

Ward 3 had been represented by African American John Morrow since 1978. A pioneer of his generation in Gainesville, Morrow had helped lead the 1977 campaign for black voting representation that expanded the City Council from three seats to five. The following year, he was elected the first black council member. Because the mayor's post rotates between council members, Morrow had been elected mayor in 1985. In 1990, frustrated at city unresponsiveness to their concerns about toxics and other issues, residents of Ward 3 sought a change.

Johnson's 1990 run for the City Council garnered a two-to-one majority of votes in Ward 3 but she lost to Morrow in the citywide at-large voting by a vote of 78 percent to 22 percent. Johnson and three others filed suit. Their aim: to bring true district elections to Gainesville and get rid of the discriminatory at-large system in which the only candidate who can represent the black community is one supported by white voters.

In surrounding Hall County, the first African American woman commissioner, Frances Meadows, took office only after a legal challenge that forced the county to adopt a district system. Meadows was elected in 1992, in a district that is 39 percent black, in part due to Voting Rights Task Force efforts. But the challenge to the city's at-large system was denied by Judge William O'Kelley after an August, 1994 trial.

Though Morrow, who was seventy-eight, passed away in October, 1996, and was replaced by Myrtle Figueras, who had no opposition and becomes the first African American woman on the Council, the lawsuit seeking the black community's right to determine who represents them remains in the courts.

"We're gonna make it"

While the work to win the political tools for local change proceeds, bringing national attention to the fight against toxics has been essential. Beginning in 1993, the Club began inviting local and state government leaders, regional and national environmental officials and others on "Toxic Tours" of Newtown. Based on the toxic profile, the tours began with a walk down Desota Street, then a bus trip through the industrial and residential sections, north and south.

That spring, with the help of the Center for Democratic Renewal, the Club invited the Racial Justice Working Group of the National Council of Churches, industry executives, and city and state officials to Newtown to see for themselves. Tours brought officials in a position to take action: city manager Al Crace, assistant city manager Carlyle Cox, state Environmental Protection Division Director Harold Reheis and state Sen. Jane Hemmer. U.S. Rep. Nathan Deal attended one briefing at the park on Desota Street.

On one tour, Rose Johnson tells about a resident with lupus who works two jobs. It is a "painful story about illness and the impact on him as a young man trying to earn a living for his family, which becomes very difficult." And, she continues, "as a result of the illnesses affecting this age range we are now feeling the economic impact that environmental contamination has on us, because people who are ill can no longer work or they work as they suffer."

Deland Wilmont had a garden behind his Desota Street home, until fear of toxics grew.

Visitors learn that few Desota Street residents still garden, a habit many miss. "Every time I think about that, it makes me sick," said Roland Waller. "I had a pretty garden. I had some fine tomatoes." He stopped gardening, Waller explained, "after I talked to this chemist. He told me, 'Well, you're safe right now, but I wouldn't eat nothing out of there. I wouldn't have no garden, because that chemical will work itself through the water and eventually in whatever you plant.' Naw, I won't plant a garden no more."

Dr. Bowen agrees "that it is not wise to garden in that community because certain plants do tend to concentrate the poisons. Newtown's runoff water and soil certainly is contaminated. The land is so sick it won't support food for human consumption."

The Club had been successful in gaining media coverage in the Gainesville paper and The Atlanta Constitution. In April 1994, CNN cameras came to Newtown to document the medical investigation for "Healthworks." A month later, Newtown residents Bush and Johnson joined two dozen other advocates across the nation to attend a U. S. Justice Department meeting to advise the department on guidelines for prosecuting environmental justice claims. While the meeting was not designed to address specific cases, "they emphasized some of the things we need to go ahead with," says Bush. "I actually felt encouraged," says Johnson.

"The future generation should be warned"

Nurturing a younger cadre of leaders in the course of the fight remains a crucial part of the Newtown Florist Club's role. A busload of forty, including many young people, traveled from Gainesville to the landmark Environmental Justice Conference in New Orleans in December 1992. They are motivated by concern for relatives and older neighbors, and by friends getting sick.

Marsha Browner "had never been sick," but nearly died the week of her senior prom. The cause of her illness has not been determined. Now, after a kidney transplant, she's active again in the singing group "Sisters." "I know somehow, I know somewhere, we're gonna make it," are the lyrics from the gospel song the group sang at the New Orleans conference. Motivated by her own experience to work for others, Browner helps Bush on environmental work whenever she can. She and other young people have formed Black Teens United for a Future (BTUFF) and attended youth leadership training to get ideas about how they can work for change in school and in the community.

What do Newtown residents want to see happen? "I think Gainesville is run with old money. I don't think that the companies will stop coming in," said one Desota Street neighbor.

"I want to see all plants now that come into Gainesville go by the safety rules for their employees, the safe equipment working 'round hazardous chemicals," said Roland Waller, before he died. "Tell 'em what kind of chemicals they working 'round and what will happen to them if they don't handle them properly. That's what I'd like to see happen. The future generation should be warned anytime they go to a plant with hazardous chemicals that they are endangering their lives."

Pat Moorehead interrupts tutoring her ten-year-old son to share her views. "I think that they will continue letting these companies move in our community. I'd like to go where it's some clean fresh air and some nice green grass, where I can sit out in my back yard and not look at garbage and smell foul odors and watch [industrial] plants continue just growing up in my back yard. I'm just afraid to stay in a neighborhood where people seem to think that something is in the air or companies are putting out chemicals that might cause me harm or my son harm."

Knowing Moorehead's concerns were on other people's minds, too, the Club initiated a meeting in the spring of 1994 to discuss relocation. Relocation "is so complex," says Lois Gibbs, executive director of Citizens Clearinghouse on Hazardous Waste, an organization that has aided thirteen communities displaced due to toxic pollution in the United States. "Relocation is probably one of the biggest dividers of the community if not approached by community leaders in a holistic way, because not everybody wants to be relocated. We must value those personal decisions."

As discussion about relocation began, other residents came forward. Johnny Hood was laid off March 4, 1994, from his job of twenty-two years as a weaver at Johnson & Johnson's Chicopee Mills. He and his wife, Jackie, grew up nearby, on Athens Highway and Atlanta Street, and bought a home on Cloverdale seven years ago. The study by Dr. Dowd and University of Georgia students identified high lead levels in the soil eight or nine feet beneath the yard where his children play basketball.

"I was kind of upset because I was under the impression that the city was aware of selling us this house and knowing it was on a dump." The Hoods were "renting through the city and they gave us the option to buy the house. ... I just didn't think it was right, and if I had a knew [about the dump] at the time I would have never moved over here. And I've been sick, and I didn't get sick until I moved over here," Hood says. Two of his children have skin rashes.

"And you know, as a black person, you can't help but feel like that you are a second-class citizen, because don't nothing go that away," he says, pointing away from Newtown to the north. "All the industries and smokestacks, everything come this way. We're virtually just stuck, we ain't got nowhere to go. And I've got a house and sittin' on a dump, and what I'm supposed to do? Sit here on it? It wasn't my mistake. I'm not trying to beat nobody out of anything. I worked all my life."

Diane Waller grew up in Newtown, moved away, and later moved back. Another of the group of relatively new homeowners at a community meeting, she expressed concerns about the older people. Some of them won't make it through a move, she says. Others are concerned about the loss of family ties if the community were to be moved. "There would be no family home," says Barbara Jones, Wilkins's daughter, who came home to Newtown for the third Toxic Tour on Earth Day 1994.

"A Victory is Overdue"

Whatever residents decide, polluters must face tighter limits. "Maybe the factories – which I know it's too much money and they are not going to move – the factories have to have stricter guidelines on their pollution," says Mize. "Because the air don't just settle here, it blows on, and other people might be probably infected too 'cause it's according to which way the wind blow," says Mize. "We, in Gainesville, not just Newtown, don't need any more toxic factories. Everybody here is going to be polluted."

There is growing recognition that "the scope of our problem far exceeds the boundaries of this neighborhood," says Johnson, "when you look at the level of contamination that is on the south side." Bush's niece, Hazel Johnson, lived in Newtown until 1956, when she moved to Black Drive, on the other side of the factories. "It affects all the south side, spitting out all the stuff," she says.

Residents have made links with other communities seeking environmental justice. Being in the Newtown Florist Club "gives me opportunities to even go further than Gainesville," says Wilmont, who recently attended a conference with Browner, Annette Westbrook, and Pauletta Wilkins. "We went on a workshop in Louisiana, a workshop on telling you how to organize and how to fundraise, or strategic thinking. I'm very fond of the Club and the only reason that I wasn't really into it before was because I worked."

Club members' courage and persistence has won respect in and beyond the immediate community. Dorothy Rucker, who lives out Route 129 on her family homeplace, the site of the historic school for African American children founded by her mother Beulah Rucker, speaks with admiration about the Club, "They were the lone group that didn't mind tackling anything. They weren't scared of nothing." Rucker continues, "They laid the foundation for our town as far as tracking down your history and not being afraid of going from here to Atlanta and Atlanta to Washington."

"The Newtown Florist Club has set the pace, all of the organizations since Newtown sprang from their shoulders," says Rucker. "Sometimes as people get older, they fade out, but as they got older, they got stronger."

"Right now," adds Whelchel, "it makes me feel better to know that some of our people understand what I went through and are trying to make it better for the next generation."

Almost a decade into this latest phase of Newtown's fight for justice, much work lies ahead: gaining more power in local government, continuing to identify the sources of toxics, pursuing health treatments for each affected person, and directly challenging companies that pollute. Government agencies must be confronted, allies expanded and tough choices faced. The fear remains that documenting the health problems, winning air monitoring, establishing the LEPC, taking Cargill to court, may improve emergency planning for all of Gainesville, but not fundamentally alter exposures in Newtown.

"We are overdue for a victory," Commissioner Frances Meadows said at the community meeting about the lupus study in 1996, capturing what many feel.

At the beauty shop on Desota Street, where a good bit of organizing goes on between clients, Faye Bush leans away from the phone to resolve a detail about the location of one potential toxic site. Despite the Club's accomplishments and the leadership role the Club has played in focusing Gainesville's attention on toxics, Bush offers a sober assessment: "We know more about it, but not that much has been done" by authorities to eliminate hazards and improve community health.

Still, as the neighbors go forward, they are fueled by the faith and action of strong women and men who will not be silenced until the community they have nurtured for the past half century is safeguarded. That spirit endures in people such as Geraldine Collins, who lives around the corner on Harvey Street from Faye Bush and Mozetta Whelchel and whose mother and sister both died of cancer. Collins, a certified nursing assistant for twenty years, lost much of her voice to throat cancer. But like the others, she is there, speaking out, "The little words I say, if that helps people to come out and try to do something about the situation, I can get up and do it again."

Special Thanks and Acknowledgments

Action For A Clean Environment
ACLU of Georgia
Atlanta Black United Fund
Citizens Awareness Group
Citizens Clearinghouse For Hazardous Waste
ECO-Action
Emory University School of Public Health
Environmental Justice Resource Center, Clark Atlanta University
Fund for Southern Communities
Georgia Environmental Policy Institute
Georgia Government Documentation Project, Georgia State University
Georgia Legal Services
Greensboro Justice Fund
Institute for Policy Studies
Lake Watch and Lillian Hall
Lawyers' Committee for Civil Rights Under Law
Legal Environmental Assistance Foundation
Manischevitz Foundation
Mary Reynolds Babcock Foundation
Bert and Mary Meyer Foundation
Morehouse School of Medicine
National Council of Churches Racial Justice Working Group
Northeast Georgia Black Leadership Council
Peace Development Fund
Physicians for Social Responsibility
Positive Action Committee
Poverty and Race Research Action Council
Public Welfare Foundation
Sapelo Foundation
Southern Organizing Committee For Economic and Social Justice
Southern Regional Council
Southern Regional Economic Justice Network
Threshold Foundation
Toni Lee and Sarah Reed
Turner Foundation
United Church of Christ-Commission For Racial Justice
United Methodist Church Commission On Religion and Race
United Methodist Women
University of Georgia-Advanced Hydrology Class
Urban Rural Mission of The World Council of Churches
Vanderbilt University Medical Center STEP Intern Program
World Council of Churches

Dana Alston
Ed Arnold
David Bailey, Esq.
Jack Beckford
Tena Bledsoe
Dr. Frank Bove
Dr. Beth Bowen
Bill Brooksher
Dr. Robert Bullard
John Clark, Esq. and Willie Woodruff, Esq.
John Egerton
Gary Flowers
Laurie Fowler
Dr. Howard Frumkin
Lois Gibbs
Wendy Glassbrenner
Tonya Gonzalez
Professor Grover Hankins
Hilary Harp, Esq.
Jane Hemmer
Tom Henderson, Esq.
Mickey Higginbotham, Rick Lavender, Velekeyta Redding
Dr. Bob Holmes
Beni Ivey
Tarik Kardestuncer
Cliff Kuhn
Charles Lee
Daniel Levitas
Worth Long
Dr. Ed Mayhew
Alan McGregor
Congresswoman Cynthia McKinney
Sybil McRay
Selena Mendy, Esq.
Sheila O'Farrell
Rodney Shanks
Damu Smith
Brian Spears, Esq.
Jerry Thomas, Esq.
Connie Tucker
Allen Tullos
Pat Wehner
Carol Williams
Cynthia Williams

To order copies of The Newtown Story: One Community's Fight For Environmental Justice, write to
The Newtown Florist Club, 1067 Desota Street, Gainesville, GA 30501
Proceeds from The Newtown Story benefit work for environmental justice in and around Newtown.